# AMAZON ECHO

*UPDATED EDITION!—COMPLETE BLUEPRINT USER GUIDE FOR AMAZON ECHO, AMAZON DOT, AMAZON TAP, AND AMAZON ALEXA*

**BY: SCOTT BAKER**

## Legal Notice:

## Disclaimer Notice:

# TABLE OF CONTENTS

# INTRODUCTION

Amazon has done an amazing job working on a variety of products that many people use to enhance their lives. The Amazon Kindle was widely popular, allowing people to take hundreds of books on the go with them in an easy format that didn't take up so much space. Since then, they have widened their base to include many other products that have become household names around the United States.

The Amazon Echo is another of these great inventions that is changing the way that people live their lives. The Echo is basically a speaker that is able to interact with the user in order to listen to commands and do the work. If the user wants to turn on music, check the weather, listen to an audiobook, or even control other smart devices in their home, the Amazon Echo can be set up to make all of this happen.

This guidebook is the ultimate guide to everything you need to know about the Amazon Echo. We spend some time talking about the basics of this device, such as what it is and how it works, before moving on to some of the steps that you need to know in order to get the most out of this device. From setting up the basic commands, linking the Echo to devices in your home to give commands to, and so much more, you will find that life can be so much easier when you use the Amazon Echo.

When you are considering getting the Echo to make your home life easier or you already have one and want to learn all the great things that you are able to do with the Amazon Echo, make sure to check out this device to get started.

# CHAPTER ONE

# WHAT IS AMAZON ECHO?

Amazon Echo is a speaker developed by Amazon. It is a cylinder speaker that has seven pieces of microphone inside and will respond to the user's voice. When the user is ready to turn on the Echo, they will simply need to say "Alexa, " and the speaker will start listening to their voice. The user can make changes to the name, choosing to go with "Echo" or "Amazon" as the wake words instead.

There is so much that this device is able to do. It can interact with voices (speaking to the person as they give commands to the speak), plays music back, helps to make to-do lists, streams podcasts, sets alarms, provides real-time information such as traffic and weather, and even plays chosen audiobooks.

Many people have fallen in love with using Amazon Echo because it is an easy device to set up, won't take up a lot of space, and makes life so much easier. They can talk with the device to turn it on, and the device is capable of recording voices while you get work done around the home.

Reports state that Amazon started working on this product in 2010 as an expansion on their successful Kindle e-readers. The Echo wasn't advertised until more recently, with a prominent feature in Amazon's first Super Bowl ad during 2016.

In the beginning, Amazon was interested in seeing how well this device would do so they offered it just to certain members. Some shoppers were invited to try out the Echo, and those who had Amazon Prime were allowed to give the speaker a try. Finally, in the summer of 2015, Amazon released the Echo in the United States and in other countries during 2016.

The coolest development for the Amazon Echo is the Alexa voice service. When you turn on the Echo and start talking to the speaker, the Alexa voice service does all the work. The voice is able to respond back to you, keep track of your lists and more. In addition, this service can be added to some of your other Amazon devices, and you can easily connect these devices with your Echo and the Alexa service.

**How the Amazon Echo Works?**

The Echo product is connected with "Alexa" which is the internet indexing company owned and operated by Amazon. This helps it to perform the different activities that you will use with the system. The device is going to stay in the default mode of listening to your speech so that it can wait for your wake word to

purchases, you can link your Echo to make reservations with Uber and even order with some of your favorite restaurants. You will find that this can make both your personal and professional life so much easier.

- Link the lights in your home—with a few basic replacements, you will be able to make your lights smart lights and will link all of them to Alexa. This can be really helpful because you can turn on or off certain lights in the home at a time, or choose to turn all of them off or on. This can be nice if you are worried that you left a few lights on downstairs at night and you don't want to go all the way down to check. Depending on the lights that you picked, you will be able to dim or brighten the lights as needed.

- Lock the doors or unlock them at specific times—you can use your Amazon Echo to take control of many parts of your home. If you replace your locks to be smart locks, it is possible to link the system with Alexa so that you can give commands on when to lock and unlock the doors. This can help the doors to be locked when you are not home, or at night even if you forget, and yet the doors will be unlocked at a specific time that you set so you can just walk into the home without having to look for a key.

- Order products just using your voice—you can even shop on your Echo. The Echo is going to already be set up with your Amazon account so you can have the device look up items that you would like to purchase. Then, as long as you have a billing address and your credit card on file, you will have the capabilities to order products just by telling Alexa that you want the product. This can save a lot of time when you're in a hurry.

- Start the coffee pot and other appliances when you want—how would it feel to have a fresh pot of coffee ready for you right away in the morning? If you have a smart appliance, you can link it with your Alexa app and easily tell Alexa when to turn on the coffee pot, slow cooker, and more to get things moving, even when you are busy with other things.

These are just a few of the skills that you can set up with your Echo and command the Alexa software to do for you. There are many more skills that Amazon is always adding on and soon it will be possible to add any of your apps to the Echo, making your life so much easier!

**What is Amazon Alexa?**

When using your Amazon Echo, you are going to rely on the Alexa software quite a bit. This is the voice service that is behind the Amazon Echo, and it is able to provide the skills that make it easy for customers to interact with the Echo and other devices just by using their voices. You can use your voice in order to set the alarm, set a timer, ask general questions, and even play music. The Alexa technology makes this easier than ever so you can get the most out of your Amazon Echo.

The Alexa software is used on many Amazon products and makes it easier to get things done when you need. Consumers are impressed with Alexa, an Amazon produced product because it makes it easier than ever to get things done with the Amazon Echo, and other Amazon products, simply by using your own voice commands.

## Software Updates

Amazon is one of the leaders in technology and is always working to provide some of the best to their customers. The Echo has gone through a few evolutions thanks to the new software that Amazon releases for it. Most of these have been to fix some of the bugs in the Echo system and sometimes the software updates will enhance how well the speakers are able to perform the tasks you want.

Since the Echo is based on the cloud, it is easy to get the new updates that are needed. In fact, many of these enhancements can be done on the Echo without any updates to the software that is already on it. This means that you can update an older Echo with some of the newer technology without having to change up the software. For example, in April of 2015, Echo added a new update that allowed live sports scores to be read out without having to change the software version running on the product.

In most cases, Amazon tries to push on these new updates slowly so it may take a few weeks before they get onto your device after Amazon announces them. This it to ensure that the devices still work properly and that your experience is not interrupted at all.

**Related Devices**

There are several different devices that were released for the Amazon Echo. Each one is a different size and be used for a variety of purposes. If you are limited on size or you would like to have a certain functionality, you will need to pick the one that is right for your needs. All of these use devices will still use the Alexa software for communication and to provide you the services that you need.

Some of the options that you can pick with Amazon Echo include:

*Amazon Echo*

This is the flagship of the company and has been one of the biggest sellers. It is easy to use and allows you to do the many special features that are listed in this guidebook. You can use it at home, at work, or even to help yourself get fit through a training program of your choice. It is all there waiting for you to get the most use out of it possible. Coming in at nine inches and with the ability to connect to all parts of your home, you are going to fall in love with the Amazon Echo in no time.

*Amazon Echo Dot*

In March of 2016, Amazon unveiled a new option called the Amazon Echo Dot. This is basically just like the Echo, but it is the size of a hockey puck. It does not come with its own speakers due to the small size, but it does hook up to your existing speakers to give commands and hear information, music, books, and more.

The Echo Dot is designed to work just like the original Echo, it just comes in a smaller size. This makes it easier to take the Echo Dot on the road with you and use it anywhere that you have a speaker.

Many people like the Echo Dot because it has all the features of the regular Echo but it is a much smaller size, and the price is more affordable. Keep in mind that because of the smaller size, the speakers are not as good on the Dot as they are on the Echo and you may need to rely on another speaker system when purchasing the Dot.

*HW Release*

As mentioned before, Amazon is always working to make changes to their technology that will benefit the customer and in October of 2016, they released a new version of the Amazon Dot. The benefits of this new Dot is that the price will be more affordable and the technology has been updated to recognize voices better. You will be able to choose between two colors, white and black and you can use the new Echo Spatial Perception technology to hook up several of the Dot and Echo units together and only have one of the devices answer you.

*Amazon Tap*

Amazon Tap is another option that you can choose. This is basically the Echo, only smaller and easier to take on the go with you, but it comes with dual stereo speakers. The Tap will have the same features as the Echo, but it is powered by a battery so you can take it everywhere. You will not need to use a password like you do with the Echo, but instead, you will just need to push a button and then speak your commands.

### Using more than one device

You are not limited to using just one of these devices in your home at a time. In fact, there are often specials that go on with Amazon that promotes purchasing several of these items at the same time, so setting them up in your home can be an easy process.

You may have as many of the Echo devices through your home as you would like to make things easier. If you would like to have one in your kitchen and in the bedrooms, that is easy to set up. If you just want to have one or two in the home, that is fine as well. These devices can be connected together in the home so that you can have as many as you would like and keep your home connected and working well no matter where you are.

Consider having a different name of the devices, especially if they are in rooms that are close together. This ensures that the right device hears your commands and that you don't get two of them trying to do the same work or competing with each other. You can name one Alexa and the other one Amazon. If you have one device upstairs and one downstairs, though, you may be fine with naming both devices the same thing since they are far enough away from each other.

## Concerns with Privacy

There have been some concerns about privacy that have come up with Amazon Echo. Since the Echo is able to listen to your conversations at all times so that it catches the awake word, unless you manually turn off the machine, many worry that Amazon is able to hold onto these conversations and will use them against the consumer at any time.

Amazon has responded to this concern saying that Echo is only going to stream recordings after the wake word has been activated. If you have not said the wake word, the machine is not going to record the words at all, even though it is still listening in order to recognize the wake word.

In addition, Echo does use some of your past recordings. These recordings are sent to the cloud in order to help improve how it responds to the user later on. The user can go in and delete their voice recordings from the cloud, but they should be aware that this can make the machine less effective at answering your questions. You simply need to go to the "Manage My Device" page and delete them or talk to the customer service at Amazon.

Echo will also use your address to help keep track of traffic updates, weather, and other important information that will relate to you. You can use the app service to set up your location. If you are planning on asking where popular restaurants are, asking about traffic, or looking at popular places to visit in your area, the mapping function can be helpful, but some people worry that this could cause issues with privacy.

Overall, there are some concerns about the information that the Amazon Echo is able to collect from regular use of the machine, such as phrases you say and your location. But Amazon has addressed these issues and is working to keep the information as safe as possible for all users.

There are so many things that you can do with your Amazon Echo. You can use it to check the weather and road conditions. You can set it up with many other programs to stream your music, news, or audiobooks. You can ask questions, set up your to-do list, and so much more. It is amazing all the things that you can do with such a simple program from the comfort of your own home!

# CHAPTER TWO

# SETTING UP THE ECHO

**O**nce you have purchased your Amazon Echo, it is time to get started on getting everything set up. This process is simple and will just take a few minutes to get the machine hooked up and to get the Alexa software placed onto it.

The first thing that you need to do is take the power cord and connect it to the bottom of the machine. If you are able to, find a location that is about ten inches away from anything like windows or walls, that would obstruct the signal. Once you have picked the right location, plug the Echo in and wait to see the LED light turn on. If it is blue, wait a few minutes because you want the light to be orange. Once the light turns orange, the Echo will greet you.

**Buttons on your Echo**

When you look at the machine, you will notice that there are two buttons right at the top of the cylinder. One will be the microphone button, and the other one will be the action button. By pressing the microphone button just once, the Echo will turn to MUTE, and you will see a red light turn on. Press it one more time and the microphone will then be turned on.

For the second button, there are a few different activities that you can do. With the action button, you can set up the Wi-Fi mode by holding in this button for five seconds, turn the set alarm off or on, and wake up the device.

Now take a look at the base of the Echo cylinder, right above where the wire is connecting into the device, you will notice there is a LED bulb. This light is going to be white when your Echo is connected to Wi-Fi. If the Echo is not connected to the Wi-Fi, this light will be orange. If you notice that the light is orange, but it is blinking, the Echo is able to get on the Wi-Fi connection, but it is not able to access Alexa.

**Downloading Alexa**

You will want to work on downloading the Alexa app at this point. The easiest way to do this is to type alexa.amazon.com and search to download it for either iOS, Android, or Fire OS. You will then be able to sign into the account and follow the instructions there on how to install the app and get the website all setup to start. If you want, you can go into either the Amazon, Apple, or Google Play stores and do this as well, but the website above can make things easier.

Alexa is a tool that can be used well with many of the Amazon products. That being said, note that Kindle Fire 1st and 2nd Generations are not able to use the Alexa app.

**Setting up your Echo**

Before you get started with setting up the Echo, make sure that you connect to a dual-band Wi-Fi network. You will not be able to do this well with a mobile hotspot so do it at home. peer-to-peer and ad-hoc networks are not going to work that well either. You need a strong internet connection so keep that in mind when starting. When you are ready to start, follow these simple steps:

- Open up the Alexa App and click on Settings.

- Select the Echo device that you have and click on the Update Wi-Fi button.

- If you are first time user, you can click on Select a new device.

- Press the Action button on your Echo device

- You will notice that this light is going to turn orange and a list of networks should appear on the mobile device. Pick the Wi-Fi network you would like to attach to and type in the password if needed.

- If you notice that the network you want is not visible, scroll down and click on Add a Network.

- This should bring up the network that you want, but if it doesn't work, you should be able to click on Rescan and get it to work.

- Now you can click on Connect, and the Alexa app should be ready to use on your Amazon Echo.

**Decoding your LED Light**

You will notice that there are different colors that will come up on your LED light. This is basically a way for you to know what the machine is doing at different times. Some of the colors that you may see and what the Echo is doing include:

- When your Echo is first starting up the machine, you will notice that a cyan light is on a blue background.

- If the Echo is processing a request that you send through and it is busy, you will see a cyan light with a blue background that goes the way of the person who is speaking.

- If the Echo is connecting to Wi-Fi, you will see an orange light that is revolving clockwise.

- If the Echo is on MUTE, there will be a red light.

- When your Echo is adjusting the volume setting, there will be a white light.

- When your Echo is trying to detect the Wi-Fi, and there is some kind of error present, you will see an oscillating purple light.

- If the Echo is waiting to hear your requests, there will be no lights present.

**Volume Ring**

Adjusting the volume on the Echo can make your experience a little bit better. If you turn the knob clockwise, you will be able to turn the Echo up so that you are able to hear the voice a bit better. If the machine is a bit too loud for you, you should turn the Echo anticlockwise.

Connect to Bluetooth

You will be able to connect your Amazon Echo to your Bluetooth, as well as connecting it to your tablet or smartphone. The first step will be turning on the Bluetooth for your mobile device and make sure that the smartphone and tablet are in the range of your Echo. Then you just need to use these following commands to make sure that the Echo knows what it should connect with:

- "Alexa, pair Bluetooth."

- "Alexa, disconnect my phone or tablet."

- "Alexa, connect my phone or tablet."

**Talking to your Echo**

Now that your Echo is set up, it is time to get started talking with it. Many people worry that this process is going to be difficult, but Amazon has worked hard to make things as easy as possible. You basically need to say the wake word to the machine, and then the Echo will be able to start listening to your commands. Sometimes it may take a bit of time to help the Echo to recognize some of your commands, but for the most part, once you set your wake word, you will be able to enjoy a seamless process with your Echo.

To begin talking to the Echo, you will need to use the wake word. The default wake word is Alexa. So you would simply say something like "Alexa, play some music" or "Alexa, what are some local restaurants." When your voice gets to the Echo, the blue LED lights will turn on, and that shows that the Echo is listening. Once the Echo has time to analyze your request or command, it will replay.

Alexa is set up to listen all the time for the wake word through the seven microphones. This makes it easier for the Echo to respond quickly once you start giving a command, as long as you use your wake word first. The seven microphones can make it easier because you won't have to really raise your voice or yell if you are trying to give a command with a lot of different conversations or when there is music on. Alexa is even able to pick up on different voices and can separate out the questions easily before getting the answers.

Alexa is the great program that makes it all happen. As long as you have your accounts set up with the Echo and you remember to use your wake word, this software will make sure that the Echo is able to respond properly to your commands.

**Changing the Wake Word**

It is possible to change the wake word if you would like to use something other than Alexa. You will simply need to do this on your computer or your smartphone. The steps to get this done include:

- Open up the Alexa app and go over to the control panel.

- You will see on the left side that there is a list and you can select the settings for which Echo device you want to make a change to.

- When you click on here, you will see which Wake Word is listed for this device. You simply need to click on this and then select that you want to make a new wake word.

- Click save and start using the new word when you want to talk to the Echo.

Remember that this change is something that you are only able to do on the Dot and the Echo. The two other options that you can make include Echo and Amazon. You can either go with these options because you like it better or because you have more than one device and want to have them both with different names to avoid confusion.

Since you are able to have a few of these devices around your home at a time, perhaps having one upstairs in the home and one downstairs to make things easier, it is best to have different names for each device. This can be hard to remember in the beginning but will ensure that the right device is activated when you are ready to use it.

**Using a Remote**

Your Amazon Echo is going to come with a remote that you can use to make things easier. This remote is going to have Playback controls, a talk button, and even a microphone. If you want to talk, you just need to press the talk button on the remote and then begin talking. If you want to pause, play, change the volume or switch to different songs or parts of the book, you can use the playback button on the remote. It is all there at your fingertips.

The main advantage of using the remote is that you won't need to use the wake word all of the time to wake up Alexa. You can just push the buttons on the remote and get the same thing to occur as you would with just talking.

**Setting up voice purchasing**

It is possible to set up your Amazon Echo to purchase physical and digital products from Amazon using their popular 1-click payment method. You will first need to set up your account online with a valid payment method and your billing address in the United States. If you are ordering physical products, you will need to set up a Prime Membership as well.

When you go through the process of registering your Alexa device, you will find that the Voice Purchasing is already on the device by default. There are several commands that you can use to make purchases on your Echo including:

- **Purchase** a Prime Item

- **Reorder** an item

- **Track** the status of an item that was recently shipped

- **Add** items to your cart

- **Cancel** your order right away after you placed the order.

Simply use your wake word and list the command that you would like the Echo to perform with your purchase and it will all be taking care of.

If you wish to make some changes to this voice purchasing power, you simply need to open up the Alexa App, tap on the settings, and check on voice purchasing. You can enable or disable the voice purchasing part, check your billing address and payment method, and make other changes to your account.

There are some times when you need to be careful with using the voice purchasing. While it can be nice and is a great convenience when you are trying to make a purchase while being busy, if you have younger children or teenagers, you may want to turn it off so they can't place orders without your permission.

There are also a few categories, mostly in physical products, that you can't purchase with voice purchasing. These include:

- Amazon Fresh

- Jewelry

- Prime Now

- Watches

- Shoes

- Prime Pantry

- Apparel

- Add-on Items

As you can see, there are many different things that you are able to do with your new Amazon Echo. These steps are meant to help you get some of the basics set up with the device so that you can get it to work for you and make your life easier. There are still so many other things that you can do for your device, and with a little bit of attention and the ability to work with the Echo, you will be amazed at what this device is able to do for you.

# CHAPTER THREE

# ALEXA COMMANDS

There are really so many things that you are able to do with the amazing software that comes with Alexa. You can basically send a command or request for anything, and it will try to help you out as much as possible. You can use one or two-word commands or go with a full sentence depending on your needs. The Alexa software is so intuitive that it will be able to help you out no matter what.

Some of the basic commands that you can use with the Echo include:

- Alexa, help

- Alexa turn up

- Alexa turn down

- Alexa louder

- Alexa, cancel

- Alexa, mute

- Alexa volume (pick a number between zero and ten)

- Alexa, stop

**Getting Help from Alexa**

Not only is the Echo great at answering your questions such as how is the weather and what the traffic is like in your area, it can actually help you if you're having an issue with the product. You can ask the Echo almost anything, including what it is able to do, what are some new features, and more.

Any time that you have a question about Echo or how the Alexa software works, you just need to ask the product. It is set up to help explain some of the issues or updates and features that it has and this can make it easier for you to use the product on a daily basis. Some of the things that you can ask the Echo to help you out with include:

- What is Connected Home?

- What are your new features?

- What is an Alexa skill?

- How can I use the Alexa skills?

- What is Voice Cast?

- What can you do?

If you have a question and want to save time looking it up, just ask the Echo, and you will find that Alexa will be able to answer most of these questions for you.

**Using the Alexa SKILL Commands**

When your Echo comes in, you will notice that it has a set of abilities that are already built in. You are able to enhance the abilities that are already on the Echo or even add new abilities to the device. You simply need to add in the new Alexa Skills when they come out, and these will be found on the Alexa App.

You can think of the Alexa Skills like the Android and iOS Apps for a smartphone. This is a newer program, so there aren't as many skills as some other devices, but the platform is really popular, and Amazon is working hard to make this a bigger thing.

It is really basic to get a skill to work on your Echo. You simply need to tell the Echo, "Alexa, launch {skill name that you want to use}. The skill will then be launched, and the Echo will welcome the name of that skill. You may also hear a message that contains information about this skill as well as some commands that will help make it easier.

Some of the skill commands that you need to keep track of include:

- The first command that you will use is "Stop." This one is pretty easy, just say Alexa, Stop and the Echo will stop. You can even do this while the Echo is speaking, you just need to say it loud enough so that the Echo will hear you over their own voice.

- Next is help. If you are looking to get some help with a particular skill or you have a question about it, you just need to say Alexa, Help and it will take you to the right file so that you can learn more about the skill. This can save you a lot of time and ensures that you are getting the right information to properly use and get the most out of a particular file.

- Resetting the Echo—you will be able to see the rest button on the Echo right at the bottom. Look near where you would plug in the power cord. You can use a paper clip or another small item to press the reset button for five seconds to help the Echo restart.

- Muting the Echo—if you want to make sure that Alexa has stopped listening to you, the best option is to press the MUTE button that is located on the top of your Echo. The light will turn red once the Echo has turned off. You simply need to press this button again if you want the Echo to start listening again.

- Enforce updates—for the most part, Alexa is going to update on its own where there are options available. But there are times when the release of an update and your Echo getting this update will be a bit delayed. You can force this update to occur a bit earlier if you would like. To do this, just place the Echo on Mute for at least 30 minutes, and the update will occur.

- Asking for calculations—it is possible to set Alexa up to ask for calculations if you need it. You can ask for basic math problems like adding, subtracting, multiplying, and dividing. You can ask for conversions between feet to centimeters and even for the temperature to make things easier for you.

- Get news updates—if you are cooking dinner or getting ready in the morning and looking to get updated on all the news of the day, you can ask for flash briefings. Alexa is set up to work with NPR, TMZ, BBC, and more. You can just as Alexa for your Flash Briefing, and Echo will show your selected sources.

  o To set up your news updates, you need to go into the Alexa app and tap on the left navigation panels. You can go to settings and select the flash briefing.

Customize your briefing with the headlines, news, shows, and more that you want to follow.

- Sports scores—you can use your Echo to update you on scores from live schedules and matches of any team that you would like. You can just ask Alexa for the score of the team you are following.

- Localized information—if you live in the United States, you will be able to get localized information. You can go into the settings on Echo App, tap on Device Location, enter the zip code of your area, and save the changes. This will make it easier to get local news, weather, and even information on pre-recorded shows depending on your area.

- Configuring updates on traffic information—for those times that you need to travel, it is nice to use the Echo traffic information to plan out your route. To make sure that you are getting the most efficient routes, you can go to the settings on the Echo App, tap on change address and put in an address in both the To and From spots and save your changes. This will help you to get the best information about traffic for your desired route.

- Reading Kindle Books—if you want to let the Echo read out your eBooks, you just need to give the command.

Make sure to list the title if you have more than one on your account to avoid confusion. You can do the same things with your audiobooks.

Amazon Alexa makes it easy to do many of the things that you want from the comfort of your own home. You will find that you can ask a lot of common questions, even some to help you answer questions about how to use the Echo device, to make life a bit easier. Once you get the practice of using the Echo, you will find that you will be asking it questions all the time.

## CHAPTER FOUR

# CREATING A SMART HOME USING AMAZON ECHO

The Amazon Echo device is not just something that you will use to listen to some music and do a few conversions for you. While these are some of the tricks that you can do with your Echo, there is so much more. For example, you can use the Echo to help manage your whole home. Set up the Echo to control the thermostat, take over all of the Wi-Fi devices in your home that can connect with the Echo, hue lights, and so much more.

The possibilities are almost endless with what you are able to do with the Echo, as long as you set it up in your home and make sure that the other parts are hooked up to Wi-Fi as well. Then with a few simple commands, you can make your home into a smart home and with a few simple commands, you can make a lot happen thanks to the Echo.

Even if you are not at home, the Alexa will be able to help you out. You will be able to use your tablet or smartphone in order to send commands to Alexa, as long as these devices are connected with the Echo, in order to get things done while you are gone. For example, you could lock and unlock your doors, make sure that the lights are off, and even close the garage door.

So what all are you able to do with your Amazon Echo thanks to Alexa, in order to make your home a smart home? Some of the options include:

- Locking and unlocking doors. You will need to have the Smart Danalock recipes to do this because it helps to give you some control over the doors so that you can auto-unlock or lock the door at certain times, making it easier to keep the door closed at night. You can use the Echo, Tap, or Dot to control the feature.

- Use the D-Link Smart Plugs in order to add some of your other smart devices to the Echo. You can add smart devices to your home at any time and easily add them to the Echo to make commands easier.

- GE Appliances Refrigerator Channel—you can even control your fridge from on the road. This channel is going to allow you to do a few different things. You can check the fridge door light if it is left open for too long or you can have the Echo set the fridge so that it is on Sabbath mode.

- The Weather Channel—this one is really neat if you are always on the run during the morning. You can use it to set your coffeemaker to turn on at a certain time, or at sunrise, or you can turn on the Smart Plug when it becomes sunset. This app will also allow you to turn on the air conditioning if the humidity or temperature rises in your home.

- Netatmo Welcome—this app is going to help you to identify different visitors that come to your home, and you are able to change the behavior for each one. For example, if a certain person shows up, you can switch the Smart plug either on or off, and if someone you don't know shows up, you can turn on the D-Link of the Smart Plug.

Get notifications on the phone if there are any motions in the home when you aren't there.

As you can see, there are a lot of different things that you can do with Alexa to make your home easier to use and to keep it protected. Before you are able to get these hacks to work, you need to make sure that you connect a few devices to Alexa and then learn how to use the IFTTT Recipes and Alexa Skills to get the devices to work the way that you want.

## Connecting Devices to the Echo

Now, if you want to set up the different parts of your home with the Echo, they need to be smart devices. Your Echo is not able to hook up with a regular light switch and turn it on and off. Rather, the lights need to be smart lights, ones that are connected to your Wi-Fi. If the lights are already connected to the smartphone in your home, they are already a part of the Wi-Fi, and you will just need to make some changes to get the Echo and the different parts of your home linked up.

There are many different parts in your home that you will be able to change into smart devices and then hook them up to your Echo. It often depends on how much you would like to change in your home and how much you would like to be able to control with a few commands. You can change your lights, the garage door, the locks on your doors, and even some of your appliances. Think of how cool it can be to just talk to Alexa to set timers, unlock the door when you come home, open the garage door and so much more.

If you want to build up your smart home, you will need to choose which devices that you would like to connect to Alexa in order to control them with the Amazon Echo. If you are looking for the best devices to use with Alexa, use some of these options:

- Lighting and fans—Haiku Wi-Fi Ceiling fans

- Outlets and switches—TP-Link with a smart plug with energy monitoring, D-Link Wi-Fi.

- Thermostats—Sensi Wi-Fi Programmable Thermostat, Ecobee3 Smarter Wi-Fi Thermostat, and Nest Learning Thermostat

- Locks—Danalock and Garageio

- Car Control—automatic

If you are using one of these devices to connect to Alexa, you will not need to use a bridge or a hub to get them connected, making things easier and avoiding issues down the line. If you are using another option, you may need to add in a hub or a bridge to help make it work with Alexa.

When you are ready to connect the smart devices with Alexa, you will simply need to go to the main page of the app and create an account. You can then connect the smart device with your Amazon account. You can then use your personal voice commands or use the Alexa App to gain control of this smart device.

**IFTTT and Alexa**

If you are looking for an easy way to link different functions and apps with Alexa, IFTTT is one of the best options. There are a lot of commands that you can do and it works well with Alexa so that you can carry out a lot of tasks without all the time and effort. But you will need to get the IFTTT connected with your Amazon account before being able to start. To do this:

- If you don't have an account with IFTTT, go to their page and set one up.

- Go to the home page for your channels and select the Amazon Alexa channel

- You will then be prompted to enter your information to sign into the Amazon account.

- Once you are signed in, you will be able to access all your existing commands for Alexa.

- There will be over 800 recipes here, and you can start adding them to your account depending on which ones you need.

There is so much that you can do when IFTTT is attached with your Alexa. For example, you can use the Nest Thermometer in order to control the temperature of your home. You just need to choose the phrase that you will use for this command and the temperature and then just state this to Alexa. But first, you will need to connect the Nest channel to Alexa through the IFTTT interface. To do this:

- Open up the Alexa channel with your computer or smartphone.

- Scroll down and choose the Smart Home option

- You will then come to the Device Links tab, and from here you can select Nest and click Continue.

- Use your Nest ID and password to log in.

- Now you will see "Discover Devices." If your Nest account is on your local Wi-Fi network, your Echo will find it.

At this point, you will be able to set a phrase that will change the temperature using something such as "Alexa, set room temperature to 76 degrees now."

**Using the SIGNUL Beacon Channel**

This is a great way to really get your Amazon Echo to work for you. This channel is going to be able to perform certain tasks automatically based on the presence or absence of your smartphone, and you can even set these to happen at certain times of day. Think of how easy it would be to just come home and have the lights turn on for you, the air conditioner or heater on, or even have Alexa turn off your phone after a certain time of day so you can just sit back and relax. All of this is possible when you set up your Echo with the SIGNUL Beacon channel.

You will need to define your zone entry and the exit events and plug these into the SIGNUL Beacon so that these tasks will become automated. The channel will then be able to use your physical presence to make all of your digital tools more streamlined. You can set it up so that the channel will do certain tasks, such as turning on the lights or change the temperature at certain times of the day or whenever it notices that you are back home.

**Grouping your lights**

It is also possible to group your lights together so that they all turn off together when you need. If you leave lights on at night outside, you can just be in your room without having to get up and try to remember all the lights that you have left on.

The trick here is to get all the lights turned on to your Wi-Fi network because Echo will be able to connect to all of these devices over the internet. Amazon only has a few types of lights and switches, but you can use other options and connect them through one of the Smart Hubs that are available.

You will then be able to use Wink or GE Link bulbs (these are some of the easiest and best to use) to group the lights and control all the bulbs in your home, but you need to get it connected to the Echo to get started. In order to get started with grouping the lights follow these steps:

- Open up the Echo app and click on your settings.

- You can find "Connected Home."

- Add Wink. When you add Wink, you will be able to see which devices are connected.

- Add these to your group.

Now you can just use the commands to get Alexa to turn off the lights. You can do one set of lights at a time, such as "Alexa, turn on the bathroom lights." Or you can group them together and say "Alexa, turn off all the lights." It is that simple to get control over the lights in your home.

In addition to setting the lights so that they will go on and off at certain times of the day. WeMo is going to make this process easier, and since it is going to run on your Wi-Fi, you will be able to keep the lights running for a long time without needing batteries.

To get started, you can download the WeMo app from iOS or Google Play and get it on your smart phone. You will then be able to hook up WeMo with any of the different devices that you would like to have connected. You can choose to do this with the lights to turn on outside ones when it gets dark out, set it up with your coffeemaker, so it starts making you a fresh pot of coffee in the morning and so much more.

As you can see, there are so many things that you can do with the Alexa app when it comes to your home. You are able to set it to take control of your home, making it easy to turn on and off appliances with just a simple command. You get to be in full control no matter where you are in the home and if you set it up correctly, you will be able to get your home to perform certain actions at different times of the day, regardless of whether you gave the command.

**Smart Appliances**

It is even possible to hook your appliances to your Echo to make life easier. You will need to get the right appliances, and most smart appliances are going to be more expensive compared to their counterparts, but for the time and convenience, you are getting, some people feel that this extra expense is so worth it.

There are many appliances that you can get like this. GE has a smart fridge that will help keep the door closed to preserve energy, and some will even hook to your smartphone to show you what you are missing in the fridge. You can find slow cookers that will turn on and off with some commands. Even your coffeemaker can be hooked up to turn on at a certain time in the morning so that your coffee is ready as soon as you are.

If you want to hook these smart appliances up to your Echo, you will need to go to the Alexa app. Look at your settings and then find the skill that you want that corresponds with the product that you want to hook up. If you already have it hooked to the Wi-Fi, you can now link the Echo with the appliance of your choice. Once that is done, it is easier to turn on and off the device as you would like, sent timers, and send the Echo some other commands that are related to this appliance.

**Locking and Unlocking your doors**

Keeping your home safe from others who may want to get inside and cause some mischief is important for your best safety. Sometimes it is hard to remember to lock the doors when you are running around and getting the family out the door in the morning, and it is a pain to try and find the right key when you get home late at night.

If you make some changes to the locking system in your home, you could let Alexa help you out with this. Once the Echo is attached to the smart locks in your home, you will be able to tell Alexa to lock or unlock the doors whenever you need. This can be helpful t make sure that the doors are locked when you leave the house in the morning or when you need to get in at night. And if you happen to get in bed and can't remember whether you closed and locked the doors at night, you can just talk to Alexa and give a command to lock the doors. It is that simple.

In some cases, you may be able to get the doors to lock and unlock at certain times during the day on a timer. This can make things easier as you won't have to have Alexa hooked up in order to give the command. For example, you could set it up for the home to be unlocked when you are scheduled to be home so that you won't have to fumble for keys or have the doors automatically lock at a certain time of night when you are usually asleep. Don't worry; you can always command Alexa to unlock the doors if you need to get back out again.

**Taking care of the garage doors**

If you are looking into replacing your garage doors, you may want to consider going with a smart garage door. This can make things easier since you will be able to hook it up to the Echo in order to set timers, keep the door locked, and even get it ready for when you plan to come home. While these are not as well-known as some of the other smart devices, it can really help to make things easier.

Depending on the service that you use with your garage door, you may be able to use the Echo directly in order to hook up the garage doors, or you may have to use a hub. Either way, you will be able to link it easily with the Echo by going onto Alexa and adding it into your skills. After a little search, the Alexa software should be able to find the garage door device, as long as it is on the Wi-Fi, and can link right to i.

The Amazon Echo is going to make your life so much easier. You will find that it is easy to connect your whole smart house to your Alexa and get it to all work just by a simple command. You will need to make sure that you are connecting the different apps to your Echo before getting started, but once you are done with this, it is easy to control all the parts of your home in seconds.

**Ordering food online**

If you are looking to find a way to order a meal for your family, the Amazon Echo is able to help you out. First, you can ask Alexa for information on what local restaurants are near you for picking a great place to take the whole family. Alexa will need to be hooked up to your location in order for Alexa to have any idea what restaurants are in your area, but this is a simple process you can do when you set up the Echo.

Any time that you want to take your family out, or if you are looking for a great place to take the in-laws, take out an important business client, or to hold a big party, you should bring out Alexa to help out. Simply ask Alexa for a list of restaurants in your area, or even split it up into the type of restaurant that you would like to visit. Alexa will be able to provide you with this information to make things easier.

If you would like to order food in to have a nice family night, Amazon Echo has even made this possible. Amazon Echo has teamed up with Domino's pizza so that you are able to order your family night pizza from the comfort of your home. Now you will have to do a few things to get this setup, but after that, the process becomes easier, so you can get pizza to your home when you need it.

First, you will need to go to the Alexa app page and add the Dominos skill to your Alexa. You will also need to have a Dominos account and have it set up with your address and billing information. To make the process work, you also need to have a quick order set up with Dominos. Set it up to have the type of pizza or the order that you would like Alexa to place for you in the "Easy Order" through the Dominos website.

After you set up your Easy Order, you will need to make sure that you send the right phrase over to Alexa. This is one of the processes that Alexa is picky on the wording with so you will need to be careful. If you don't say the words properly, you will find that Alexa gets confused and you will not be able to get your pizza order.

According to Amazon, if you want to order your pizza from Dominos, you will need to use the command "Alexa, open Dominos and place my Easy Order." While this may be a larger chunk of a command compared to some of the other options, it will help the Alexa to open up the Dominos app that you need to get the pizza ordered. Over time, this will also help to avoid confusion when Amazon will hopefully add some more ordering options to the Alexa. It may seem pointless now since Dominos is the only option to do this, but Amazon is always working to improve their features, and it won't take long to make this process a bit more streamlined.

After you have placed your order, you will even be able to track your Dominos order to see how far it is from your home. You will just need to say the command "Alexa, ask Dominos to track my order." Alexa will then be able to tell you where your order is located and let you know how long it will take to get the order to your home.

While Dominos is the only online food ordering service through Amazon right now, this is hopefully something that Amazon will improve in the future. Many families would enjoy a few simple steps in order to set up their food orders and get things to their front door in a few seconds. And for now, it is easy to find other local restaurants in your area so that you and the family can have a fun night out together.

The Amazon Echo has many great features that will make your life so much easier. Whether you want to be able to create links to take control of your smart home with a few simple commands or even order in some food for the family to enjoy, the Amazon Echo has all the tools that you need to make life easier.

# CHAPTER FIVE

# GOING TO WORK WITH YOUR AMAZON ECHO

**N**ot only can the Amazon Echo make it easier to control your home and do basic tasks around the house, but it can also help you out with getting things done at work. You will be able to use the Echo in order to keep track of important events, organize your workflow with others in the workplace, order an Uber driver to get you from one location to another, and so much more. Why work harder at work when you can work smarter with the help of Amazon Echo and Alexa.

**Connect your calendar with Alexa**

There are many different calendars that you will be able to use with your Alexa so pick the one that you like the most to link up with the device. We will talk about how to hook up Google Calendar because this is one of the easiest ones to use. To connect with your Google Calendar, follow these easy steps:

- Open the Echo App on your phone or on the computer.

- Then click on settings, Calendar, and then Link Google Calendar Account.

- You can then provide your Google account information. The next time you activate Alexa, you will be able to check your schedule.

At this point, you are only able to check on events that you have already scheduled. If you would like to be able to write new events into the system, you will need to integrate with an IFTTT recipe called "Add Amazon Echo To-Do to Google Calendar.

Once this is integrated, you are able to add simple commands to the calendar to adds important dates and information that you need to keep track of for work and other events. Make sure to keep the commands simple and easy to understand so that Alexa can do the work for you without getting too confused. You should also avoid using the first person with Alexa, rather just avoid pronouns if you can't get rid of the first person in your talking.

Now it is time to note the events that you want with the precise time and even people right onto the calendar. To do this, open up your computer or mobile phone and connect with the Google channel. You will be able to use the Slack channel to make this process easier so that everyone member of the family, or your team at work, will be able to add their personal information in.

You can also do this by using Amazon Alexa. Just go to the app and go on to your settings. Click on the Calendar and then pen in the new event that you want Alexa to remember.

If you would like to remember repeated tasks that you need to complete often, you need to use the Trello channel. It is going to help you to organize these repeating tasks and will also send over reminders when these repeating tasks are coming up in your schedule.

Setting up your calendar is going to make it so much easier for you to keep track of all the things that happen during your day. You will be able to tell Alexa when there is a new event and have her put the information into the calendar for you. Each day when you need information about your schedule, you will just need to ask Alexa for this information and the software will be able to check it out for you. You simply need to set up your calendar to work with Alexa, and now your whole schedule will be in one place.

**Sharing your workflow**

At times you need to share information with others at work. You may have a webpage, or a document that you need to get over to someone else through Google Docs and the Echo is going to be able to help you out. You will be able to do this with the Google Drive Channel and Workflow Channel being connected. Once the two are all connected, you will be able to share any text that you like with Google Docs.

For those who are already using Slack, you will be able to hook this up with the Workflow Channel so that you can send web pages and messages to others in your team with ease.

**Ordering Uber**

If you work in a job that is often on the go and you are always calling an Uber ride to help out, looking the number up and calling them all the time can become kind of a pain. You can order your Uber using the Amazon Echo; you just need to get it all setup. To do this just:

- Open up your Alexa App and then tap on the menu bar on the top left of the screen.

- You can then tap Skills and look for Uber.

- Click on Uber to help enable it.

- Sign into the Uber Account (or create a new one if you don't already have an account) and then click on Allow.

You will then be able to ask Alex to call for an Uber right with you. It takes just a few seconds while you are working on other tasks and you can have a car waiting for you in just a few minutes whenever you need. This can make it easy to call a car ahead of time while you are getting ready to leave without having to wait on the phone all the time to get the information. You can pack up to leave, finish the meeting, or send out that email at the same time while talking to Alexa about ordering your Uber car.

**Finding the right restaurants**

There are times when you will need to schedule a lunch or dinner with a client. This can help to provide a more relaxed atmosphere and helps you to convince the client that you are the right person to work with. Whether you are starting out as a smaller business or you have been around for a while, it is a good idea to have a list of different restaurants that would be perfect for different situations with your clients.

Even if you don't usually take your clients out to dinner, it is still a good idea to have these local eating places on hand. This can make it easier to bring in breakfast as a treat, order in lunch if you haven't gotten all your work done, and so much more.

The Amazon Echo will be able to help you get all of this done in no time. If you set up your personal location already, you will simply need to ask Alexa where some local restaurants are. You can just be general and ask where these restaurants are located, or you can look for a specific type of restaurant, such as Italian, pizza, or another option.

When you put in your command, Alexa will get to work looking at which restaurants in are in your local area that meets your criteria and give you information about the ones that you are the most interested in. Pick out one of your favorites or try something new and you are set to impress the clients in no time or to treat your employees when needed.

And if you are interested in sending out for pizza to have a party for your team members at work, or just because everyone needs to work late to get things done for a big project, you can work with Dominos to get this done. Simply add the Dominos app to your skills, set up your address, billing information, name and your easy order, and then you can just tell Alexa to place the order for you anytime you need.

**Setting up your DocSend Channel**

This can be a great option to use if you need to keep track of all the documents that you are sending throughout the day. You will be able to connect the different recipes together to determine when someone has gotten ahold of your document and even when they have completed reading the document. You can also send a follow-up email after someone has read the document to check that they got the update or if they have any questions. Some of the channels that you can use to do these skills include:

- Gmail—this is the one that is the most popular and can make it easy to connect with the DocSend channel. Any time that you have a visitor to the group, you will receive a notification.

- ORBneXt—with this option, you will receive a notification when a visitor comes on or even when someone read through the whole document. This can help you to keep track of whether everyone on the team has had a chance to read through the document and you can remind those who haven't had a chance that they need to get it done.

- If Channel—this is another one that allows you to tell when someone is on the document with you and who has had a chance to read through the whole thing.

Sending documents to others on your team with the help of Alexa can make things so much easier. You will be able to interact with each other, ask questions, and keep track of the progress of each other. It is simple to handle and can help to improve communication between you and the other people working with you on a project.

**Setting Up Square Channel**

This is a great tool to use if you accept online payments. If you work from home or have another business that could do work online, you should set your Echo up to work with the Square Channel. You simply need to go and set up your own Square account before you can activate it on your Echo, but once you do, you will be able to enjoy the many features including:

- Payments—receive an email anytime that money has been sent to your Square account so that you can keep track of which customers have paid you and which ones are still owed.

- Refunds—any time that someone uses your Square to make a new refund, it is going to add in another line to the spreadsheet. You can even receive an email when any of your accounts receives a refund of a certain amount so that you can keep track of the money you are spending.

- Funds are going into an account—any time that there are funds going to your account, whether it is a new payment or a refund or some other thing, you can receive an email for your account.

You can use Square with a number of other channels including Gmail, ORBneXt, and DocSend. Just make sure that you have the Square Channel set up and that you are sending all of your business transactions through here in the future.

**Adding in your blogging content**

Are you someone who likes to spend time blogging? Is this a way that you help to promote your business across the world, or are you more interested in following a certain blog in the hopes of keeping up to date on the things that matter the most in your line of work. The Amazon Echo can make it easier to add these blogs to your device. You can use one or several channels at a time using channels like Tumblr, WordPress, and Blogger. You can even use some other options, such as Salesforce, Quip, Slack, or LinkedIn to add in your own personal business network.

Picking the blogging network that you want to use for your work life or for your Echo to keep track of can be a challenge. Some people choose to go with many of these but depending on your business; this could end up being a ton of extras on the Echo. Some of the best blogging channels that you should consider to keep up to date with your work include:

- Tumblr—this is a great site to use if you would like to publish a lot of photos. It can work with Flickr and Instagram so you can merge all of your information together and post a blog all in one place for your customers to take a look at.

- Blogger—Blogger is great if you would like to integrate a ton of other things in with it. You can use it to blog any images you have on Dropbox and even share these new posts on Facebook to easily reach your consumers and let them know when new information is posted. If you are interested in doing Vimeo vids, these are easy to add to Blogger as well.

- WordPress—WordPress is a great place to start if you are a beginner with blogging. It is easy to use and really stable, and you can make things even better when combining with echo. You can use WordPress with Tumblr, add YouTube videos when needed, add your pictures from Instagram, and even publish your blogs online all thanks to the Echo.

There are other options when it comes to creating a log with the Echo; you simply need to choose the blogging site that works the best for you and then set up your account before sending over the information to your Alexa account. Once you do this, it is easy to tell Alexa what you want to get done with a few simple commands.

Blogging can make a big difference in your business and can propel it forward. With good content, you can bring in more advertisers who would work with you to place content on your page. You can sell different products to the customers who come and visit you and get ranked higher in search engines. Why not spend some time learning how to work with the Alexa software so that you can get your blogs published and running in no time?

**Speeding up the complex actions**

If you are doing some work that needs some complex actions, you may be confused on what you should tell Alexa in order to get it done. This can be common when trying to get multiple steps done with your work. Using the Launch Center Pro will be able to save you time and headaches by allowing you to complete these complex actions with just one command.

To do this, you are going to need the Waze app. To do this, you will need to sign up for the application and then set your Launch Center Pro so that it starts to get notifications. Now you can connect the Launch Center channel in and take control of getting these actions sped up, even if they would normally take a few commands to complete.

Using the Amazon Echo is an easy process in your daily life. Whether you are at home or you need to put this speaker to use when you are at work, you are going to learn so much using this device. You can get so much done thanks to the Echo, using it to order an Uber car, find the right place to eat for a business meeting, sending documents and working to make sure that everyone is able to collaborate together. No matter what you need to get done at work, you will find that the Amazon Echo will be able to help you out.

## CHAPTER SIX

# USING YOUR AMAZON ECHO TO GET FIT

**W**e have already discussed many of the great possibilities that you can do with your Amazon Echo. You can use it to give basic commands around the house, such as listing out the news and playing some of your favorite music. You can also use it to turn your home into a smart home, setting up the lights, locks, and other appliances so that they are attached to the Echo, and you can control them with a few simple commands. You can even bring the Echo to work with you and use this as a way to get more done in the workplace from sharing workloads and documents to ordering in food and calling over a car if you need to travel.

In addition to all the other great things that you are able to with the Echo, you can also use it to help you get in shape. Everyone is looking for ways to make getting in shape a bit easier. From tracking how much exercise, you are doing, figuring out how fast your metabolism is going, and even how much you eat during the day, staying on a diet can seem like a hopeless endeavor without some help. And now you can add in the Echo to make sure you are meeting all of your health and fitness needs.

The voice control on your Echo is going to make getting fit easier than ever. While you can use many other tools to help you stay active and be as fit as possible, nothing is as effective on its own as it can be with the Echo. You can add in your exercise equipment and even your Fitbit to the Echo and get even better results than before.

Depending on the type of exercise equipment you are using, you will be able to attach it to the Echo. This can be helpful in several ways. You will be able to use the Echo to change information on the machine, such as the amount of weight or even the incline of the machine. If you are uncertain about what to do on the machine (such as if the machine is new or you want something different for the workout), you will be able to talk to Alexa to learn something new.

You can also use the Alexa feature on your Echo in order to keep up with many of the health trends that are going on. New York Times is a great place to get this information and can be easily linked in with the device. Add this to your Echo and start to get emails sent to you with tips and tricks for the best health and exercise routines.

You can use the fitness part of your Echo in order to turn on the Withings app to keep track of your body measurements. You simply need to add in the Google Drive Channel and the Withings Channel so that you can record the measurements and have them stored to keep track of your measurements. If you like to keep track of your sleep schedule and the amount of steps that you take, it is a good idea to download the Fitbit Channel and have this information sent to Google Drive.

As you can see, there are so many things that you are able to do with the Echo that can improve how well you keep your health up. You can choose to keep track of your body measurements, sleeping times, steps, and even learn some new tips and tricks to keep you moving throughout the day. These are simple to set up; you simply need to open up IFTTT and look for the recipes that are listed there.

**Connect the Echo with Fitbit**

Fitbit has quickly become one of the best fitness tools on the market. There are so many options available that can help you keep track of your daily activity. Many people who feel that they are healthy and active are surprised when they put on a Fitbit and find out that they aren't moving that much or aren't burning as many (estimated) calories as they had assumed.

The Fitbit is a great way to track many parts of your health from your sleep schedule, how many steps you are taking, how long you are working out, your resting heart rate, and more. For many people, it is a great way to motivate themselves to get up and get moving, making it easier to stay healthy and lose some weight.

You can choose to link up your Fitbit with your Amazon Echo in order to update and keep track of your daily fitness goals. While this is still one of the areas that Amazon needs to work on because the interaction isn't as smooth as some other programs, the Fit Assist will take a look at your goals and give you some facts and tips about fitness and health.

Alexa is not going to store the data about your daily activities with the Fitbit (you can download the Fitbit app to keep track of this if you wish), but you will be able to use the information that Fit Assist sends you in order to make new goals each day. Just add this part on by going to alexa.amazon.com and looking under skills.

If you would like to connect your Fitbit to the Amazon Echo, you will need to do so through the IFTTT. You will need to head over to the Alexa website and connect both the Google Calendar channel and the Fitbit channel. Make sure that your goals are set on the Fitbit so that Alexa can send you the right reminders.

Once both of these channels are set up and linked with your Echo, Alexa will be able to help you adjust your sleeping schedule, tells you when it is best to go to sleep, and will provide you with a spreadsheet of your activity through Google so that you can keep track of everything. This can be a nice way to see your progress and get reminders of how to get enough sleep to lower stress and to keep functioning through the day.

**Skills available through Alexa for Exercise**

For those who are all into the fitness and health, or even for those who are just getting into it all and want to make sure they are on the right path to the healthiest life possible, there are a lot of great apps that you can use with your Echo to make things easier. Some of the options that you may want to try out include:

- FitnessLogger—when this is used along with Alexa, it is going to help you to record your specific fitness schedule. It doesn't matter what fitness schedule you are on, or even if you are trying to mix it all up, this will help you to record the exercise and then compare workouts throughout the week. You simply need to say "Alexa, ask FitnessLogger for all supported exercises" and then you will receive a list of all your exercises. This helps you to see if you're staying consistent or if you need to pick it up a bit.

- 7-minute workout—this is going to help you to cut out the fat in your life and even to lower stress. You don't need to have an hour of working out each day to see results. Sometimes a few minutes is enough to make a big difference. Or maybe you just had a stressful day and just need a few minutes to reduce the stress. Seven minutes of a quick workout with the help of Alexa can make all the difference. All you need to say is "Alexa, start a seven-minute workout, and you are ready to go.

- Training tips—if you are new to working out, make sure to set up the training tips. This can help you to learn the best workouts for your needs, and you will be able to get the most out of your gym time. You just need to ask Alexa for tips for the day, and you are ready to go.

- Recon Channel—this app is a bit more expensive than the others, but it is top of the line and will boost how well your Echo is able to take care of your workouts. It is meant to track your fitness by projecting your metrics to an eyepiece. You will be able to see how well you are doing on the workout while moving, without having to stop to check out the statistics. You can also get sports news, DubNaiton updates, calendar updates, and more from here once you get it set up to the Echo.

- ALOP-Pilates-Class-Skill—if you are interested in learning how to do Pilates as a good start to your workout or as a way to get up and stretch a bit more for sore and aching joints, this Pilates app is the best one for you. You can simply add this into your Alexa skills and then tell Alexa that you want to start the Pilates class. This skill will take you through a whole exercise schedule to help you get something new each day. If you would like to set it up, just check out ALotOfPilates.com.

You can also use the app to track your food intake, the different exercises that you are doing, and to keep track of the different measurements that you need to take. This can help you to get started on your new workout program or to make your current one even better. No one wants to go on a plan and find out that it won't be that effective, but with the help of your Amazon Echo, you can add in a few different apps and get the best results in no time.

Using the Amazon Echo is one of the most efficient ways to keep track of your workouts and to ensure that you are going to get the results that you want. There are apps present for those who are just starting on their workout adventure and some for those who have been at it for a long time. Use Alexa to link your favorite apps and find how to keep your body as healthy as possible.

# CHAPTER SEVEN
# STREAMING YOUR FAVORITE MUSIC ON THE ECHO

The Echo is excellent for streaming music. You no longer need to deal with setting up a CD player, getting your computer to find the right music that you need or any of the other systems that will provide music. The Echo makes it easy to set p a few of your favorite music programs and then give a simple command to get it all to start working.

There are several different music sources that you can already use for Alexa, and the list is growing every day. Alexa will work well with both free as well as subscription-based streaming services including:

- Audible

- iHeartRadio

- TuneIn

- Pandora

- Spotify Premium

- Prime Music

- Amazon Music

If you have any of these options, you will be able to download the music that you would like to listen to and get it to turn on with a few different commands. The commands are simple with this option. You can ask Alexa to play music, stop the music, skip songs backward and forward, pause, and even continue on after pausing. Just telling Alexa what you would like to do will make it easier to turn on the music that you would like to use.

Keep in mind that depending on the streaming service that you are using, you may need to change around some of the commands that you use. TuneIn and Spotify will work with different commands compared to the Prime or Amazon music options, so you will need to make some adjustments if you are streaming music from these sites to make sure that Alexa will understand you.

**Moving your music over to Amazon**

You may have to move some of your music over to your Amazon or Audible Library. It is not going to work if you have it on the Amazon Cloud Drive so if you are having trouble with getting the Echo to play some of your favorite music, you may need to make some adjustments to where the music is located. Amazon Music Library is where you should store all the music that you would like to play and Audible Library is best for any audiobooks that you would like to listen to while getting work done.

The good news is that if you are already an Amazon customer and have an account, you already have the information you need for an Amazon Music Library. You don't even need to sign up for their premium version; you will already have this account since you are a customer with Amazon.

When you are setting up the Music Library, you will be able to store a maximum of 250 songs for free. The good news is that if you are already a Prime member with Amazon, you will be able to get unlimited access to the Music Library. This means that you will be able to upload files from this library over to your personal music library and then these music files won't count towards the 250 free song limit.

Now, if you have a large personal music library that is already outside of Amazon and you want to add it over to the Amazon Music Library, you can choose to go with a premium subscription. This is less than $25 a year, and you will be able to hold up to 250,000 tracks. This amount is really affordable for those who want to hold a lot of songs in their Music Library so they can always switch around the different songs that they want to listen to. And just like the other option, if you make a song purchase from Amazon, it is not going to count towards your limit.

The choice of which subscription you would like to use is a personal one. If you don't have a ton of music to add to the Music Library, you may want to just go with the free option. But if you want to have the option of adding lots of songs onto the Echo, it is really easy and affordable to join the premium group to hold more songs. Some people choose to start out with the free plan and then migrate over to premium if they eventually add enough songs to make it worth their time to have more space.

Some people don't want to deal with the hassle to migrate all of their playlists to the Amazon Music Player. If you have lists from many sources or you have a bigger list, this may not be the choice for you. There are other options that you can use that will avoid this hassle and make it easier to get the songs that you would like even if they aren't in your Music Library.

What you will need to do is turn on the Bluetooth on your Echo and connect it directly to Google Music or iTunes. Alexa will be able to pair with any of the other devices that you need to get the playlist, as long as it is connected to Bluetooth. You will just need to use the command "Alexa, pair" and keep the device is within the range of the Echo to make it work.

Once Alexa has a chance to detect your specific device, it will give you instructions to go to the new device and go onto the Bluetooth pairing screen. You will be able to select the Echo. Once these devices are paired, you can open up the app, whichever one you use for iTunes or Google Music and start the music. You can use the same commands when Alexa is paired with these options as you would when it is paired with the Amazon Music Library.

## Accessing other music libraries

At some points, you may want to get onto other music libraries. If someone else in your family has their own music library and you would like to use that, you are able to thanks to the Amazon Echo. To do this, you need to go onto the Alexa app and set up Amazon Household Profiles for everyone in your home. This allows a second person to use your Amazon account and they will be able to do the same things, providing the same commands, as the original owners of the account. This allows everyone to put their playlists onto the same account, making it easier for everyone to share the music no matter who is home.

The biggest worry with this option is that someone else in the family, such as one of the kids, may make a purchase that you do not approve of. You may want to consider setting up a password so that no one is able to purchase anything on the account without your permission. To set up this password you need to:

- Go to your Amazon Household Profile and go to the Voice Purchasing option inside.

- Enter in a code, which will be 4 digits, that you will be asked for each time you want to make a purchase.

You should keep the code to yourself unless you want someone else to make purchases on your account without your permission. Each time you would like to make a Voice Purchase, you will need to give Alexa this code before it will process the order.

Once you get the Household Profiles set up, it will be easy for the different members of your family to change over to their profile, look up their own stuff, and even listen to their own music. You will just need to ask Alexa to switch accounts around any time that a new user is ready to use the device. If you get onto your Echo and are unsure about which account you are using, just ask Alexa and the device will be able to tell you.

**Buying music for your Echo**

You can even use your own voice in order to purchase the music you would like to use for the Echo. Simply tell Alexa to shop for the particular song you want to listen to, or even for the album name. For example, saying something like "Alexa, shop for the song (name of the song)" and Alexa will help you to find the music that you would like. You can also do this by the name of the artist if you are unsure about the name of the song.

Once you make a purchase, you will be able to store it for free on the Amazon Music Library, and it won't even count towards the storage limits. You can use them on the Echo device as well as any other Amazon device that supports Amazon Music. This makes it easy for you to download the music you want without worrying about how much storage you have left or where you will be able to listen to the music. As long as you order it on Amazon, you can store it for free and listen wherever your Amazon devices are located.

**Connecting the Echo with other music apps and devices**

Amazon Music is not the only kind of music you will be interested in listening to. If you have playlists from other sources, you will be able to connect these with your Alexa device and get a chance to listen to them with a few commands. There are a lot of music devices and apps that work well with Alexa including

Spotify Premium, Pandora, iHeartRadio, and more. You can set up a subscription to these or use the accounts that you already have to make them work for you.

First, let's take a look at how to setup and then use Pandora on the Echo. To do this, you will tap on the Alexa app and then get onto the Menu located on the left corner of the screen. Open Music and then Book before tapping on Pandora in the submenu. You will then be on the registration page for Pandora, and from here you can tap on "Link account." You can then sign in to Pandora if you have an account or you can sign up if you would like to get started.

Once you have gotten this much setup, you will be able to see the various station lists from Pandora on your Alexa app. You can then tell Alexa to play music from your Pandora list.

Next is iHeartRadio. This is another great subscription service that you can sign up for, and it includes a lot of the great classics, oldies, jazz, top 40 songs, hip hop, and even more that you would like to use. If you have an account already, you can simply go onto your Alexa app and go into settings to link iHeartRadio with your Pandora account. Then when you want to listen to this service, just talk to Alexa and let it know that you would like to listen to your iHeartRadio stations.

TuneIn is another choice that you can go with if you are looking or a wide variety of genres for the younger demographics. You can also set it up to go with some of your favorite podcasts if you are interested in listening to some of those. With this one, you will need to remember the station or the number of the dial for TuneIn to get the right station or you will need to remember the name of the podcast to get it to show up. Some sample commands that can be helpful with TuneIn include:

- "Alexa, play {number on the dial} {station name} on TuneIn"

- "Alexa, play {podcast name} on TuneIn."

You need to remember to say the name of the program that you want to use. The default is going to send you to the music that is on your Amazon Music list so if you want to use TuneIn, Pandora, or one of the other sites, you will need to state these names in your command to avoid confusion.

Some of the other programs and apps that you can add to your musical library that also work well with Alexa include:

- Fluffy Radio—this is a great one if you would like to listen to the radio while listening to Alexa. You can help you to even make requests for the music to play, such as saying "Alexa, ask Fluffy Radio to request {name of song}" and it will send in the request for you. This can make it fun to listen to the songs that you want, even on the radio.

- Spotify Premium—this is a popular option, but it does cost you a bit of money each month since a lot of the commercials are taken out. There are different packages that start at $19 a month and can go up to $80 each month.

- Tune your guitar—for those who are aspiring musicians and who are trying to get a bit better at their craft, using the tune your guitar app can help out. You can turn on the app and get the notes that you need to tune a guitar before getting out there and playing. To get the Tune Your Guitar app to work simply:

  o Open your Alexa Skills, and search around for the Guitar Tuner to set it up.

  o When you are ready to tune the guitar, simply say "Alexa, ask Guitar Tuner to tune my guitar."

  o The tuning is going to start right away. It will start from low E to high E and hit all the notes in between.

  o You can continue to tune your guitar until it is all ready to play.

There are many other apps that you can add to your Alexa program to get the music that you love to show up on your Echo. You simply need to be assigned to the right Wi-Fi network and look through the Alexa skills to find out how to link the two accounts together. You do not have to sign up for a particular music account (unless you want to) in order to start listening to music. Pick out the music site that you would like, or go with the Amazon Music Library for free music, and then do the right commands to tell the Echo to start playing.

Music is such a big part of the lives of so many people. It can help you to get work done faster, with better concentration, jam out while cleaning the home, and just put you in a better mood, even when you are getting ready for work in the morning. Using your Amazon Echo and linking it to your favorite music streaming service can ensure that you are getting all the music that you need at any time.

# CHAPTER EIGHT
# PICKING THE RIGHT AMAZON ECHO FOR YOU

The Amazon Echo has taken the world by storm. It is yet another great product that people have fallen in love with that was produced by the tech giant, Amazon. You are going to soon find that you can use this device for everything in your life from ordering your food, sending documents for work, finding restaurants in your area, running your whole house, and so much more. It won't take long before you wonder how you ran your life, your work, and your whole household without this product.

There are several options to this device, even though it is relatively new. You may be wondering which of the options is going to work the best to help you get your work done and run your life. While all of the options are amazing, you want to make sure that you pick the perfect one to fit into your life.

When it comes to using the Alexa technology and deciding which Echo is the right one for you, you have some choices. There are three main choices that you can make with this product; the Echo, the Echo Dot, and the Tap. Each of these has some amazing features that you will be able to use to really enhance your life.

When you are first looking at this Amazon product, you may be confused at which one is the right one for you. Let's take a look at each of these products to determine the differences, and the similarities, so you can make the right decision on this purchase.

**Amazon Echo**

The Amazon Echo is the first version that came out. It is 9 inches tall and works as a speaker that will recognize your voice when you give a command. It is a long cylinder that is often compared to a can of Pringles in looks. However, there is so much that this little product is able to do outside of just providing you a way to listen to music.

Those who use the Amazon Echo to its full potential will find that not only does it work as a music player, but it will easily become the controller for your whole smart home once you connect it to the different devices that you have. Whether you are using smart lights with the help of Philips Hue, a Nest thermostat, or some other product in your home, the Amazon Echo will be there to help you out.

You can think of the Echo like a search appliance for your smartphone without the screen. It will answer your questions, check the weather, check out the traffic, run your whole home like a remote, and even read audio books out loud to you. Basically, you can think of this as your new personal assistant; with a simple voice command, you are able to ask any question and even have it send documents, order for a car, and order food for you just by giving a simple Alexa command.

With the Echo, the Alexa software is always going to be connected, which is something that has concerned some people in the past. The good news is that the Alexa is not able to record anything without you saying the wake word (which will usually be Alexa or Amazon), so you will only have the machine recording you when you want to state a command.

The Echo is set up to record some of your commands, though. This is so that the device is able to get smarter and start to recognize your speech patterns, some of your preferences, and it makes the device better able to deliver based on the commands that you give.

Those who are looking for the complete package with this system will want to go with the Amazon Echo. It has a great speaker and makes it easier to control your whole smart home. It is the most effective if your home is already set up with lots of smart devices, but you can still have fun with this device even if your home is not a smart home. Whether you are asking fun questions, listening to music, using this to make reservations or get a car, or one of the other many great purposes, you will not be disappointed when it comes to the Amazon Echo.

**Amazon Echo Dot**

The Dot is basically the smaller version of the Echo. It is a lot smaller, about the size of a hockey puck, making it easier to carry around on the go if you plan on using this a lot for your work life. You will also be dealing with a much quieter speaker so it may be a bit harder to hear music and other information, but if you have the device right by you, it won't make much of a difference.

The Dot is able to do everything that the Echo does, so picking this one is not going to be a hard choice. It does have an output jack and connectivity through Bluetooth so you can connect it to a sound system if you would like it to be a bit louder.

Many people like to go with the Dot because it is smaller in size, can be hooked up easily to your current sound system, so you are going to get the best sound ever. The fact that it is only $50, which is much less than the other Echo products makes this option really affordable and appealing to people who want to have all the connectedness in their home without all the price or size.

While the Amazon Echo is often considered one of the best products in this lineup, it is sometimes seen as a bit expensive, especially for people who aren't sure if they are going to like this product or not. The Dot can be a cheaper alternative to trying out the Echo, and if you are considering getting a few of these to have around your home, the less expensive price of the Dot can make this a bit easier.

This product is best for people who already have a good audio setup and already have several smart devices in the home that you would like to set up to control with your voice. You shouldn't expect the best sound quality from your Dot, but if you are looking for something that can hook up to your sound system and perform all of the other tasks of the Echo for a fraction of the cost, the Dot is the right choice for you.

**Amazon Tap**

The second generation of the Echo is known as the Amazon Tap. It is going to bring some more of the great features that you have come to love with the Echo, but it has some added benefits, and it is really easy to move around with you. The Tap is a rechargeable option that comes with a charging cradle that is wireless. It also lets you have a maximum of nine hours of playback, so you will be able to listen to it for a long time even without the charging part.

This is the first option that is powered by a battery. In order to save you some battery life, the Tap is not always going to be running like the other Echo products. To get Alexa to listen, you will not say the wake word; rather, you will need to tap on the device to get it to work. You also need to be connected to Wi-Fi to get this to work.

Just like with the Dot and the Echo, you will be able to use the Tap to do many of the great features that you love. You can ask for music, control the devices that are in your home, and even get stuff done at work. The only difference is that you won't be able to just say the wake word; you will have to get up and tap on the machine to get it to work.

Another added bonus of going with this option is that you will be able to make some customizations to it. This is the only option in the Echo family that can be somewhat customized as you will be able to pick out a Tap Sling cover to protect the device. This Sling is currently available in six colors for you to choose from.

There are some positives and negatives that come from using this product. First, if you are looking for a speaker that is really portable and will let you stream the music you want from your favorite systems, the Tap will help you out. It also works helps you to do all of the other things that you have learned to do with the Echo and the Dot.

Some people aren't fond of the idea of having to get up to turn on the Alexa software each time they need it. Part of the appeal of using the Echo or the Dot is that you can just say a word, and the machine starts to work, but the Tap requires you to have physical contact with the machine.

For some people, this is a bonus because they feel more secure knowing that the device is not on all of the time and recording their voices. For others, it kind of defeats the ease purpose of the Echo if they have to constantly get up and touch the device to do all the tasks that they want.

# CHAPTER NINE

# EASTER EGGS AND TIPS FOR YOUR AMAZON ECHO

So far in this guidebook, we have spent a lot of time talking about a ton of things that you can do with your Amazon Echo. And while these aspects are going to make a big difference in how well you can run your life and how much fun you can have with the Echo, there are other commands that you can do when it comes to the Echo.

You do not just have to ask questions about your day or give commands that make the Echo do something that you need to get done, such as order an Uber drive or look up restaurants. You can also have some fun with the Echo. Basically, you are able to ask the Echo anything, and the Alexa software is going to answer it based on the information it has present and the skills that you have decided to use. This means that you can have a little bit of fun with the Echo and ask it some silly, or just plain fun, questions along the way.

Whether you are bored and want something fun to do or you just want to see all the great things that Amazon Echo is able to answer for you, these Easter egg phrases are going to make the Echo not only useful but so much fun. Some of the questions and commands that you can ask the Echo (although these are in no way an exhaustive list) include:

- Do you have a boyfriend?

- What is your favorite color?

- Roll a dice

- Tell me a random fact

- I think you're funny?

- How tall are you?

- Where are you from?

- Do you want to play a game?

- Who is the walrus?

- What is love?

- Do aliens exist?

- Who let the dogs out?

- Do you want to build a snowman?

- To be or not to be

- Beam me on

- Where are my keys?

- Rock paper, scissors

- What are you wearing

- What's in the name?

- How many angels can dance?

- Mac or PC?

- Tell me a joke

- Give me a hug

- Party time

- Are you lying?

- Who is your daddy?

- When is the end of the world?

- Do you like green eggs and ham?

- Do you know Siri?

- Can you give me some Money (you should ask this one twice)

- How do I get rid of a dead body?

- Is there a Santa?

- Will you marry me?

- I've fallen, and I can't get up.

These are just a few of the fun and silly commands that you can give to Alexa and see what answers you can get. They can be a lot of fun if you're looking to do something different with Alexa, or you can use them as ways to get used to Alexa and how the software works and you want to get it to start recognizing your voice. Mess around with these as well as some of the other fun options that you can come up with and see how much fun the Amazon Echo with Alexa can be.

**Animal Sounds**

In addition to some of the fun phrases that you can say to Alexa when you first get the machine, there are also some great animal sounds that you can pick from to make it fun. If you are already an Amazon Prime member, you will be able to pick out these special soundtracks and music in order to get the animal sounds, and you can get it all for free. If you are not a member, it is going to be about $0.89 for every track that you want to have.

There are a lot of different tracks that you can pick from, and many of these are really interesting. You should take some time to look at the different reviews from customers to ensure that you are making a purchase that is just what you want. You can choose from bird sounds, different animal sounds, and other mixes. Then you can ask Alexa what sounds different animals or birds make and Alexa will be able to look at the track and give you a response.

**The maker channel**

This is a gem for those who like to get into crafts or design their own products, either for the fun of it or to sell to others. You can get connected to the Maker Channel. You can work on a do it yourself project and then go on to the Maker Channel in order to share how you did the work, or at least your experience with it, with others who are interested in learning. You will simply need to use your Echo to go to huckster.io in order to share the experience with others.

As you can see, there are a lot of great things that you are able to do with your Amazon Echo. It is not just a silly little speaker that will answer some of your questions. You are able to use this to control so many parts of your life including running a smart home, helping to run your music, make purchases, answer your questions, and so much more. There is no tool like this one on the market for helping you to get things done at work and even helping you to have some fun with some of the silly questions above. It is not going to take long for you to learn how to use the Echo and soon you will be in love with all the things that it can do in your life.

# CHAPTER TEN

# YOUR SAFETY WITH THE AMAZON ECHO

There is a lot to love when it comes to using the Amazon Echo. You can use it to take control of many aspects of your life and to make things so much easier than before. It is always waiting to hear the wake word and learn what command you are sending over to help you out. Whether you want to listen to music, have a question to ask, need a reservation or a ride, or want to control your smart home, the Amazon Echo is there to help you out.

But despite all of the great parts that come with owning the Echo, there are some people who have concerns about this product. They feel that there are some issues with the safety and privacy that come with the Echo and are worried that Amazon is using this information inappropriately, either for their own advertising uses or to sell to other third parties.

But is your safety really at risk when you use the Amazon Echo? Here we will explore some of the top concerns that come with owning an Amazon Echo and what you need to now about your personal safety.

Is my Alexa spying on me?

The first concern that some people have with the Alexa system is that it is spying on them. These concerns come from the fact that Alexa is always listening to hear the wake word. People assume that because Alexa is "listening" it is catching on to other conversations, movements, and other things of the users even when they aren't using the Echo.

Your Echo is not spying on you at all. It is not going to react to you until you say the wake word. While Alexa is on all of the time, unless you mute the device or turn it off, it is not going to record your voice ever and will only do a response once you use your wake word.

For those who are worried that Alexa is using its technology in order to hear you and even record your personal conversations, it is possible to turn the device off. You simply need to push on the MUTE button and Alexa will be turned off and won't respond even if you aren't using the wake word. Just remember that you need to hold it in for about five seconds so that the ring turns red before it is off. You will also need to turn it back on before it will respond to the wake button so check to see if the red LED light is on if you can't get Alexa to respond to you.

So far, there aren't any cases of people who have used Echo to perform illegal activities to date. But there are a few precautions that you should use to ensure that no one else is able to use your Echo when you aren't there or that something doesn't go wrong with the program. Some of the things that you can do to ensure the device is always working properly include:

- Mute the Echo any time that you are not at home to prevent it being used by someone else.

- Never place the Echo close to an answering machine or a speakerphone because this can sometimes confuse the product.

- Never put the Echo near a window.

While the Echo is not designed to listen in on your private conversations and is only going to respond to you if you use your wake word, you should know that Echo will save some of your past conversations with it in order to provide improved responses later on. The Echo uses a learning system to recognize your voice and then will store these conversations so that it can make changes to how it interacts with you over time.

If you are worried about Amazon having these recordings, it is possible to delete all of these recordings, but keep in mind that you will notice that the user experience with Echo is not going to be as good. If Echo is not able to learn your commands, which is what happens when the voice recordings are saved, it is going to have a hard time working seamlessly with you in the future.

To delete these voice recordings, simply go through the following steps:

- Open your Amazon Echo app

- Tap on Settings and then on History

- At this point, you will be able to see the list of all requests that you have made to the device since you set it up.

- To delete the recordings, tap on each one and then tap delete.

- You can also delete all of the recordings by heading over to Amazon and signing in.

- Click on your devices and then choose the Echo.

- Click on Manage Voice Recordings and then delete all of the recordings if this works best for you.

**Preventing purchases on the device**

If you have more than one person who is using the device for their own personal reasons, or you have younger kids in the home who may learn how to use the Echo easily, you may be worried about the voice activated purchasing. You don't want to have your kids make purchases that you don't approve of, especially if they go onto your credit card and you may not catch them for some time.

Luckily, there is something that you can do to prevent unauthorized people from making a purchase. You simply need to go into your Amazon account and set up a password. This is going to be a four-digit code that you will have to state before a purchase can be processed. Make sure that you don't give anyone else access to your code or it defeats the purpose of having the code in the first place.

Even if you don't have anyone else in the home who would use your Echo, it is a good idea to set up this code. It can help to keep you safe in case someone gets into your home or in case someone gets ahold of your Echo and tries to make purchases. It is just four digits, so it is not going to be that hard to remember and won't be difficult to say when it is time to make a purchase.

**Troubleshooting**

There are going to be times when the Amazon Echo is not going to work the way that you would like. Sometimes the Echo is not going to listen to your commands, or there is another issue. Some of the ways that you can fix issues with the Echo listening to you include:

- Check to see if the device is on MUTE. If the light is red, the device is not listening to you. You simply need to press on the MUTE button and then start by saying your wake word to get the device to listen.

- See if the device is getting an update in software. Sometimes this can slow down how it reacts to you, and you may need to give it some time to finish the upload.

- Unplug your device and leave it like this for about 60 seconds. You can then plug the Echo back in. Sometimes this will help the Echo get reconnected with the Wi-Fi in case this is the issue.

- You may need to leave the Echo unplugged for a few hours to give it a break and to allow some updates to come through. After this time, you can re-plus the device and see how it works.

- If the Echo is still not working that great, it may be time to reset the Echo. This one is hard to do because you will need to go through the whole process of setting up the device again. But if the Echo is having a lot of issues and you just can't get anything else to work, this one often does the trick.

- If nothing else is working, even resetting the device, it is time to call in Customer Support with Amazon and see how they are able to help you.

While there are some legitimate concerns about the safety of using your Amazon Echo in terms of protecting your personal information, Amazon is a company that you can trust to protect your information and never uses it in a way that you can't trust. They have designed the Echo not to take your voice recordings and allows you to turn off the device when not in use and even delete some of the command recordings if you wish. You have nothing to worry about when it comes to your safety and your privacy when you are using the Amazon Echo in your daily life.

# CHAPTER ELEVEN

# TROUBLESHOOTING YOUR AMAZON ECHO

The Amazon Echo is a great product that you are going to want to keep around for a long time to come. It can make so many tasks easier to handle in your life, from playing music, to running your whole home (if you already have smart devices in place), to ordering food, keeping track of things at work and so much more.

For the most part, this device is going to work just the way that you would like it to. You will plug it in, download the software, and then be able to connect anything that you would like to add to your commands. Then with the right wake word and a good command, the Alexa software with Echo will be able to complete the requests that you want to be done.

But there are some times when the Echo is not going to work the way that you would like. There are some simple problems that can be really frustrating with your Echo, that you can often fix on your own. This chapter will take some time to look at common problems that come up with the Echo and the simple fixes that you can try to get the Echo to work well for yourself again.

**The Echo can't find your devices**

One of the benefits of using the Echo is that it can sign on to your smart devices through the home so that you can run them with a simple vocal command. This makes things easier without you having to run around the home to get the tasks done. You can set up your locks, your lights, your fridge, and other appliances around the home to work with your Alexa so that you just have to give some simple commands to turn them on, off, and more.

But there are times when Alexa will have trouble finding these smart home devices. If Alexa is not able to find the devices, it is really hard for the software to have any control over them and your commands aren't going to work. The first thing that you should check is whether the device is one that is supported natively in the system. If it's not, you may need to take a few more steps to get it to work with the Echo.

The list of those apps that are natively supported with the Echo is growing, and some of the devices that work well with it include Ecobee3, Philips Hue, Nest, lifx, Insteon, Honeywell, and Winx. There are many other devices that are supported with Alexa as long as you go through the Skills. This is not going to be originally recognized by Alexa though, so you will have to go through and add them in to make them work.

To add these new devices, you will just need to open up the Alexa app and go to Smart Home. Tap on Discover Devices and then go under the Your Devices part. You will be able to see here whether the devices are natively supported. If they're not, you can go through the IFTTT channel to add the devices.

Now, if you've gone through and added the devices to your Alexa account, either natively or through IFTTT, and Alexa is still not able to find them, there are a few other solutions that you can try including:

- Check which command you are using with the Echo. The commands that you are giving will make a big difference on whether the Alexa is able to understand you or not. Even small changes in the names or the phrasing of your smart devices can confuse Alexa, and it won't be able to connect. Look up what the device is called and then use this in your command.

- The issue may be more with your smart device rather than with the Alexa. Some have issues with their software that will make it hard for it to stay connected to Alexa. Checking the power cycle of your devices can help to see if there are any connectivity issues that you have. Sometimes you may need an update to the device to fix some of the software issues.

If none of this is working for you, you should try to reboot the speaker and then try to remove the device from your Alexa and then add it back in to see if this will help.

**Alexa has trouble staying connected to the Wi-Fi**

At times, your Echo device may have some issues staying connected to the internet. When this happens, it is impossible to interact with Alexa because it isn't able to record you if it is not hooked up online. The first thing that you should do is check your personal internet connection. If you are having trouble with your personal internet connection, it is time to do some troubleshooting to fix this issue or talk to your internet provider to see what is up.

If nothing is wrong with your internet, it may be time to take the next step and try out a few other things to get the Alexa to stay online. Some of the things that you can try include:

- Reboot everything—one of the easiest things that you can do with your Echo when it doesn't connect is to reboot everything. Just turn off the power and give it some time to reset. Unplugging the power adapter and then perform a power cycle on the router and modem at the same time.

Keep the Echo off until the modem and router have time to come back on. You should notice that the Echo is working much better after this time.

- Move your Echo—sometimes the issues is that the Echo device is just too far away from the power source. If you have the router in the basement and the Echo is on the second floor across the house, you may have found your issue. You should have the Echo and the router located centrally in your home. Having them higher up in the room can help avoid issues with barriers getting in the way. Keep the devices closer together as well so the internet signal doesn't have as far to travel. Try to keep the Echo away from metal objects or the wall by at least 8 inches to help with connections.

- Reset—this is not always the best option to try because it is going to get rid of some of your settings and it doesn't really let you know what the original problem is with the device. But it can help to solve almost every problem that you have with your device. It is going to set the device back to the original factory settings, so you will have a clear slate. When you are ready to try a factory reset on the Echo, follow these steps:

o Get a paper clip or another small item and place the reset button. This is at the bottom of the Echo near the power adaptor.

o You should hold this button in until you notice the top of your Echo turns orange.

o Now you have to wait for the light to turn completely off and then turn back on again.

o Now you will need to bring out your computer or smart phone and open up the application for Amazon Alexa.

o Walk through the whole setup process for the Echo again, just like you did in the beginning.

o Once you are done with these steps, the device will be back to its factory settings and ready for you to use again.

• Call customer service—if none of these tips are helping you out, it may be an issue with your hardware or the service provider. The easiest thing to do is call your interest provider and see if a spotty connection is the issue. If you work with the internet provider and that doesn't help, it is time to talk to the customer service at Amazon. Be aware that they will have you go back through all these steps just to ensure that you have tried everything. But they are the best bet when it comes to figuring out how to solve the issue.

**Alexa is having troubles hearing you**

Over time it is easier to notice that Alexa is not hearing you as well as before when you first got the program. There may be something wrong with the speakers, but the first place you should look is doing a quick reset and see if this will help you out.

The first thing that you should do is turn off the speakers and then turn them on again. Do a test run and see if this helped out and if the Echo is able to hear you properly now. If this doesn't seem to fix the issue for you, try moving the object around a bit. You need to make sure that the Echo is a minimum of 8 inches from the wall and that there are not any obstructions that are going to make it hard for the speakers to hear you.

The obstruction can be something very simple. For example, having the air conditioner on can muffle your voice and make it hard for the Alexa software to hear you. You may not notice this if you bought the device in the winter, but once summer comes and the loud air conditioner turns on, it can be hard for Alexa to help you out. Moving the device closer to you, or at least further away from the object that is bothering it, can make it easier for the Alexa to respond to your commands.

You may also need to work with the voice training app to get Alexa to understand the phrases that you are saying. With this app you will read out 25 phrases, using your typical voice and from a distance, you normally would relative from the device. This helps Alexa to start recognizing your voice so that it can respond better.

One thing to keep in mind is that if you go in and delete the voice recordings that Alexa stores after your commands, you are going to decrease how effective Alexa is. If Alexa is not able to store this information, it is not able to learn your preferences. If Alexa is not comprehending what you say and you are deleting your voice recordings, stop doing this for a few days and see if it makes a big difference.

**Accidental activation of Alexa**

There are times when the Echo will turn on without you giving a command to it. If you have a television show on or someone says a word that sounds similar to the wake word that you have chosen, you may find that Alexa is activating more often than you would like. It really isn't going to harm anything, especially since Alexa does not record your conversations at all, but it can become a nuisance if you hear Alexa responding all the time when you aren't giving a command.

There are a few basic things that you can do to help minimize this issue. These include:

- Move your device over—if it is too close to the television, it may be time to move it to another location to avoid this issue.

- Press the mute button—this will turn off Alexa, so it won't wake up, even when you say the wake word until you unmute it.

- Change the wake word—perhaps it is time to change the wake word to make things easier. Remember that you do have several choices including Alexa, Amazon, and Echo.

Amazon is still working on the voice activation features to make this a bit friendlier for the user. They are looking into teaching Alexa how to recognize your specific voice style, something that would prevent accidents happening in normal conversations or from the television, and other innovations that would make this product easier to use.

**The Echo won't turn on**

You may experience issues with the Echo not turning on when you want to use it. This will often happen because of a few simple fixes but can be frustrating in the process. Some of the reasons that the Echo won't turn on and won't respond to your commands includes:

- Not plugged in—you should take a look to see if the power cord was unplugged or if it is falling out a little bit. If you are using a power adapter, check to see if this is working properly.

- Faulty power cord—if everything is plugged in correctly, it may be time to check the power cord. Take a look at the cord to see if you can notice any damaged or frayed areas. If you do find this damage, make sure to replace the power cord as soon as possible.

- Faulty hardware—in some cases, you may have some issues with the hardware inside the Echo. If you have tried some of the other options and it is still not working, you may need to call into customer support with Amazon to see about getting replacements for the Echo hardware.

**The Echo won't connect to your Bluetooth**

Your Echo device needs to connect with Bluetooth if you would like to get it to work properly, especially if you are working with the different smart devices in your home. You may not be able to stream it to your phone or get the music that you want either. Some of the things that you should check if your Echo won't connect to Bluetooth include:

- The device is too far away—you should make sure that the Bluetooth device is no further than 30 feet from the Echo. If it gets further away than this, the Echo is going to have a hard time communicating with it.

- Bluetooth hasn't been enabled—check to see if the device you are trying to use with the Echo is actually able to support a Bluetooth connection. If it is, make sure that the settings have Bluetooth turned on.

- The device and Echo haven't been paired—you should look at your settings menu through Bluetooth to see if the device is connected with your Echo. If they aren't, you should command Alexa to pair and then choose the devices that you would like to put together.

- Bluetooth module is broken—if the Bluetooth module on the Echo is broken, you will need to get it replaced.

**Issues with sound on the Echo**

With this issue, you have turned on the Echo, and it appears to be connected, but you can't get any sounds to come out of the device. Some of the things that you should check in this instant include:

- Is the Echo muted—if you pressed the Mute button, this is going to prevent the Echo from working because it won't listen to any of your commands. Sometimes this happens without you having known it. Turn off the mute button and try turning the volume up a bit with a rotation of the volume ring clockwise and see if this works.

- Broken speakers—you need to be really careful with the speakers on the Echo. If you use the music too loudly, you may have an issue with broken speakers. If this happens, you will not only need to replace the speakers, but also the tweeter and the woofer.

**The sound is distorted on the Echo**

With this issue, you are able to listen to some of the sounds on the Echo, and it is playing and listening to your commands, but the sound just doesn't come out right. It can be too heavy with the bass, sound small or muffled, or have another issue wrong with this. If this is a problem you are dealing with, check out these areas:

- Broken tweeter—if the sound is coming out with a heavy bass to it, you may need to replace the tweeter.

- Broken woofer—if you are listening to a sound and notice that it is really high pitched, the woofer may be the part you need to replace.

For the most part, you are going to fall in love with your Amazon Echo. It is easy to use, has a lot of great features, and will last you and your family a long time with normal use. Of course, there are times when things are going to run into issues, or you will need to fix something on the Echo to make it work as good as new again. Luckily, most of these issues are easy to fix, and you can make the necessary changes from the comfort of your home without having to contact customer support or pay a lot of money to a tech person to help you out. If you are experiencing some of the issues that are described in this chapter, make sure to try out a few of these helpful hints to get your Amazon Echo back up and running in no time.

# CONCLUSION

The Amazon Echo is top of the line when it comes to Amazon products. While Amazon has been known to produce many great products, including their flagship product, the Amazon Kindle, the Echo is breaking new ground with all the great feature and how great the Alexa software is able to work on it. There are just so many things that you can do with the Amazon Echo, that it is quickly going to become a household name.

Many people assume that the Echo is a simple tool that allows them to play music and maybe ask a few questions about the weather. While you are able to do these commands, there is so much more that comes with the Echo to make your life easier. You can say simple commands like turn on the music, read a book to you, create a to-do list, and so much more. In addition, the Echo can be a great way for you to take control of your whole home; set it up with your other smart devices, and you will be able to control the whole home even when you're not there.

This guidebook has spent some time talking about the Amazon Echo and all the amazing features it is able to do. Whether you are a beginner with Amazon products and are looking to see if the Echo is the right option for you, or you already own an Echo, and you are ready to make it your own, this guidebook is here to help. Get your questions answered, learn more about the Echo, and learn how to set up each of the features that you want in one easy place!

CPSIA information can be obtained
at www.ICGtesting.com
Printed in the USA
BVOW05s0525191216
471211BV00034BA/1517/P